SURPRISING

ANIMAL WEAPONS AND DEFENSES

by Teresa Klepinger

CAPSTONE PRESS
a capstone imprint

Published by Capstone Press, an imprint of Capstone
1710 Roe Crest Drive, North Mankato, Minnesota 56003
capstonepub.com

Library of Congress Cataloging-in-Publication Data is available on the Library of Congress website.
ISBN: 9781669078401 (hardcover)
ISBN: 9781669078357 (paperback)
ISBN: 9781669078364 (ebook PDF)

Summary: Engaging text gives readers a quick rundown of unique animals that use unique and surprising weapons and defenses to survive in the wild.

Editorial Credits
Editor: Mandy Robbins; Designer: Dina Her; Media Researcher: Jo Miller; Production Specialist: Tori Abraham

Image Credits
Getty Images: Andrey Nekrasov, Cover (bottom), 24, BlackCatPhotos, 6, Brent Durand, 16, Gerald Corsi, 25, JohnCarnemolla, 14, Paul Starosta, 13, Reinhard Dirscherl, 19; Nature Picture Library: Solvin Zankl, 27; Shutterstock: Aleoks, 9, Beverly Speed, Cover (top left), Cormac Price, Cover (top right), Craig Cordier, 21, FoxGrafy, design element (throughout), Gerald Robert Fischer, 17, kaschibo, 5, Simagart, 29, Tomasz Klejdysz, 7; Superstock: Frank Lane Picture Agency, 10, 11, Minden Pictures/Ch'ien Lee, 22, Minden Pictures/D. Parer & E. Parer-Cook, 15

Printed and bound in China. 5827

TABLE OF CONTENTS

Expect the Unexpected 4

Making Zombies 6

Crazy Claws and Hooks 12

Freaky Fangs and Venom 20

Didn't See That Coming! 24

Glossary 30

Read More 31

Internet Sites 31

Index 32

About the Author 32

Words in **bold** are in the glossary.

EXPECT THE UNEXPECTED

Watch out! You might not see these animal weapons and defenses coming.

BLUE-RINGED OCTOPUS

A blue-ringed octopus blends into the sand. An unwanted visitor gets too close. Bright blue rings appear on the octopus's body. They are a warning. The octopus may bite. Its bite could kill.

MAKING ZOMBIES

PARASITIC WASP

A wasp lays an egg inside a ladybug's body. The egg hatches. The **larva** eats the ladybug's insides. But the ladybug doesn't die!

Then the larvae breaks out. It spins a **cocoon** between the ladybug's legs. There, it stays covered and safe.

PARASITIC FLATWORM

A snail eats bird poop. But it gets more than a meal. Flatworm eggs are in the poop. They hatch inside the snail. Then the young flatworms crawl into the snail's **eyestalks**.

The eyestalks swell. Birds think they look tasty. They eat them. Now, the worms grow inside the bird and lay eggs. The bird poops them out. A snail eats the poop. The circle begins again.

EMERALD JEWEL WASP

An emerald jewel wasp stings a cockroach. The roach can barely move. The wasp takes the roach to the wasp's nest. There, the wasp lays an egg on the roach and buries it. When the egg hatches, the larva eats the cockroach.

CRAZY CLAWS AND HOOKS

HAIRY FROG

A **predator** grabs a hairy frog. But the frog can easily defend itself. Claws break out of its skin and stab the animal. The frog's skin heals up when the claws pull back into place.

DUCK-BILLED PLATYPUS

Duck-billed platypuses have **venomous** claws. They stick out from the inside of their back ankles.

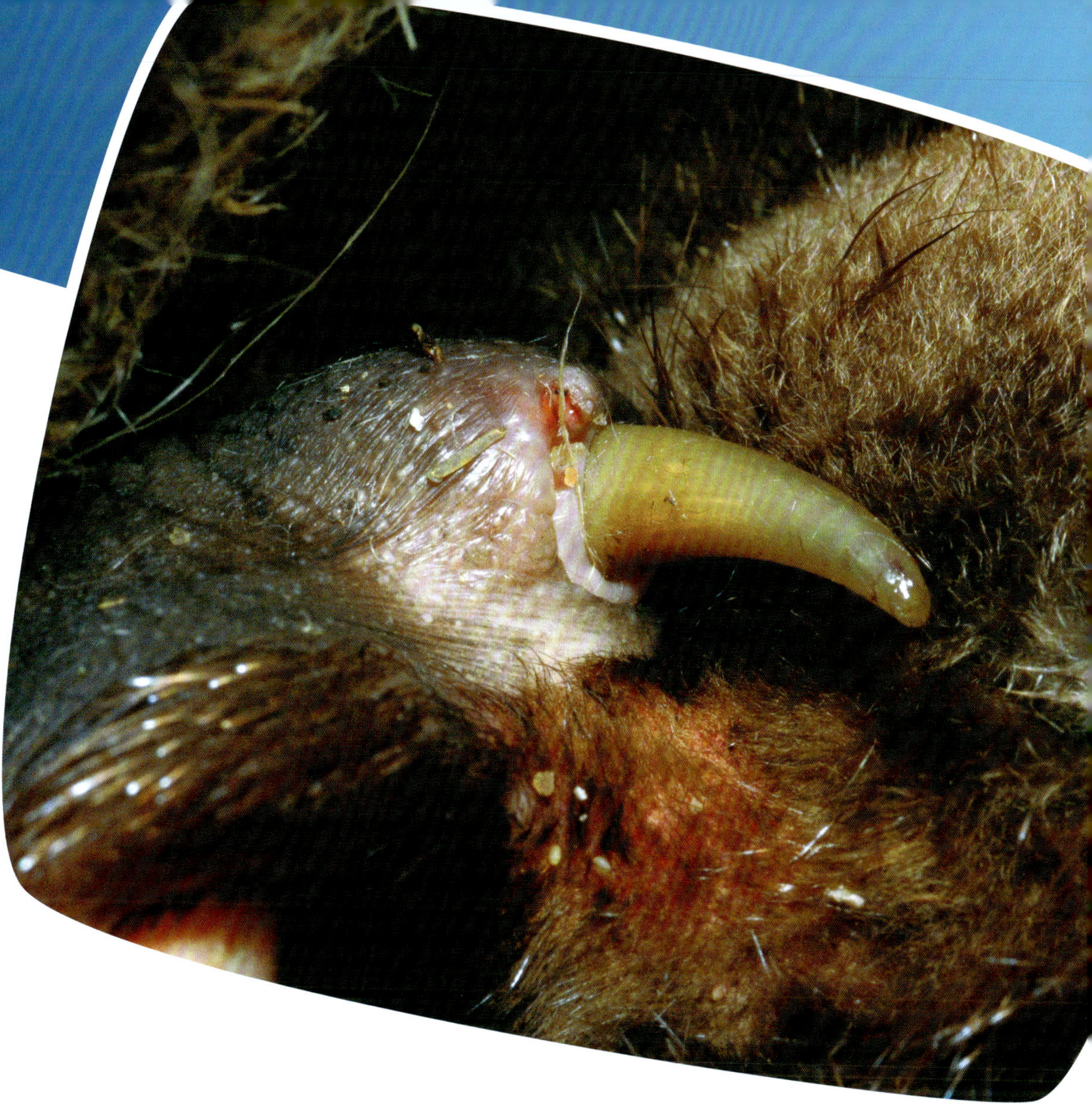

Two males fight for a mate. They kick with their back feet. These claws work like a snake's fangs. They can force painful venom into their rival.

PEACOCK MANTIS SHRIMP

A mantis shrimp creeps up on a clam. *WHAM!* It breaks the clam's shell with strong, fast blows.

Mantis shrimp have claws like clubs. They hit 50 times faster than your eye blinks. They can even break a fish tank.

GEOGRAPHIC CONE SNAIL

A cone snail smells a fish close by. In a flash, it shoots out a hook. The snail's hook sends venom into the fish. The fish can't move. A thread connects the hook to the snail. The snail pulls the fish back and swallows it.

FUN FACT

Cone snail venom can kill a human.

FREAKY FANGS AND VENOM

STILETTO SNAKE

The hungry stiletto snake chases a lizard down a hole. But it won't need to open its mouth to attack. These snakes poke their fangs out to the side. They can strike to the side and the back.

SLOW LORIS

A slow loris is ready to defend its home. It licks a **gland** near its elbow. **Toxins** from the gland mix with its spit. Now, the loris's spit has venom. It can fight other lorises with these painful bites.

DIDN'T SEE THAT COMING!

SEA CUCUMBER

A crab tries to eat a sea cucumber. But the sea cucumber has a surprise in store. It shoots its guts out of its butt. The crab gets covered in sticky threads. Then the sea cucumbers can escape. The guts grow back later.

FUN FACT

Some sea cucumbers can grow more than 6 feet (2 meters) long!

DEEP-SEA ANGLER FISH

Angler fish swim in the dark. Female angler fish have a spine in their heads. It's like a fishing pole. On the end are glowing **bacteria**. **Prey** are drawn to the light. Fish swim up to it and into the fish's mouth!

BOXER CRAB

A boxer crab wears two anemones like boxing gloves. Then a puffer fish comes by, hoping for a meal. The crab punches the fish. Ouch! The fish gets a painful sting from the anemone's **tentacles**. Boxer crabs also use these weapons to catch the small fish they eat.

FUN FACT

Boxer crabs will steal anemones from each another.

GLOSSARY

bacteria (bak-TEER-ee-uh)—very small living things that exist everywhere in nature; some are helpful, and some are harmful

cocoon (kuh-KOON)—a covering made of silky thread; insects make a cocoon to protect themselves while they grow and change

eyestalk (EYE-stahk)—a long body part that holds the eyes on some animals

gland (GLAND)—an organ in the body that makes chemicals

larva (LAR-vuh)—an insect at the stage of development between an egg and an adult; more than one larva are called larvae

predator (PRED-uh-tur)—an animal that hunts other animals for food

prey (PRAY)—an animal hunted by another animal for food

tentacle (TEN-tuh-kuhl)— a long, arm-like body part some animals use to touch, grab, or smell

toxin (TOK-sin)— a poisonous substance produced by a living thing

venom (VEN-uhm)—a poisonous liquid produced by some animals

READ MORE

Abramovitz, Melissa. *A Toad That Explodes and Other Cool Animal Facts.* North Mankato, MN: Capstone Press, 2019.

Johnson, Rebecca L. *Nature's Ninja: Animals with Spectacular Skills.* Minneapolis: Millbrook Press, 2020.

Stewart, Melissa. *Ick! Delightfully Disgusting Animal Dinners, Dwellings, and Defenses.* Washington, D.C.: National Geographic Kids, 2020.

INTERNET SITES

10 Awesome Animal Adaptations
science-sparks.com/10-awesome-animal-adaptations/

Insects and Arthropods
easyscienceforkids.com/all-about-insects/

Predators: Science Trek
tpt.pbslearningmedia.org/resource/idptv11.sci.life.oate.d4kpred/predators/

INDEX

anemones, 28, 29

blue-ringed octopuses, 4
boxer crabs, 28, 29

claws, 12, 14, 15, 16

deep-sea angler fish, 26
duck-billed platypuses, 14, 15

eggs, 6, 8, 10
emerald jewel wasps, 10

fangs, 15, 20

geographic cone snails, 18, 19

hairy frogs, 12

larvae, 6, 7, 10

parasitic flatworms, 8
parasitic wasps, 6, 7
peacock mantis shrimp, 16
puffer fish, 28

sea cucumbers, 24, 25
slow lorises, 23
stiletto snakes, 20

tentacles, 28

venom, 14, 15, 18, 19, 23

ABOUT THE AUTHOR

Teresa Klepinger lives in the Pacific Northwest where she watches coyotes hunt, looks under rocks, and pokes things with sticks. She loves any kind of animal, teaching kids, reading books, music, and, most of all, her family.